Dr. Carl Peters

# KING SOLOMON'S GOLDEN OPHIR

A RESEARCH

INTO

THE MOST ANCIENT

GOLD PRODUCTION IN HISTORY

BY

Dr. CARL PETERS.

NEGRO UNIVERSITIES PRESS
NEW YORK

Originally published in 1899
by Simpkin, Marshall, Hamilton & Co., Ltd.

Reprinted 1969 by
Negro Universities Press
A Division of Greenwood Publishing Corp.
New York

SBN 8371-1835-2

PRINTED IN UNITED STATES OF AMERICA

# PREFACE.

*AT a time when, more than ever, there is great striving to discover the yellow metal in all continents, it will be of double interest to find out in what part of the world the oldest mining country of history was situated. It is not only of historical interest but should prove of practical value, because there is no evidence that the ancient nations, with their primitive tools, really exhausted the gold mines they were working. The whereabouts of "King Solomon's Mines" does not appeal to the fancy alone.*

## *PREFACE.*

*I would not dare to publish an essay on a subject which for thousands of years has been treated over and over again, were I not certain that I could add something new to the many conjectures already brought forward; something which I think will give a clue to the unravelling of one of the oldest and greatest mysteries in the history of mankind.*

*I have much pleasure in publicly expressing my thanks to Mr. Frank Karuth, to whom I owe the following excellent translation from the German, in which language this Essay has already appeared.*

*The English edition contains a few additions to the original text.*

CARL PETERS.

# King Solomon's Golden Ophir.

**The question of Ophir.**

THE question of Ophir rises out of the obscure vastness of primitive history into our own times like an enigma of mystery. Thousands of years have brooded over the position of Solomon's Ophir, and all the time it was held, that the solution of this question would throw a bright light on the political and commercial relations, that existed between the nations on the Mediterranean and those on the Indian Ocean—on the threshold of the history of mankind.

**Carl Ritter's Opinion.**

"Out of those spaces and times," says Carl Ritter in his Geography, vol. xiv, fol. 348-9, "but few hints have been transmitted to us; the most important being the record in Holy Writ of the Hiram-Solomonic voyage to Ophir. It briefly relates a fact of such significance, that, since the time of Flavius Josephus and of the Fathers of the Church, Eusebius and Hieronymus, its elucidation has exercised the ingenuity of the most celebrated and learned investigators in the most different languages and branches of science. It has offered such vast material for numerous distinct problems, that the most divergent branches of science have profited thereby, though the main question itself still remains unsolved. It has only

ripened towards one or the other extreme probability, but proof has not yet been brought, which country must be understood by the name of Ophir. For, the further we look back towards the beginning of things, the more numerous become the possibilities for their explanation from ever so many points of view, because the sum total of positive data must decrease in the ratio of the growth of our distance from the object.

"Now, even if we may say that the entire cycle of possibilities has been exhausted in the explanation of this historical tradition; exhausted from all points of view, be it the criticism of the text, or the interpretation of the objective of the voyage, or the etymological authentication and the origin of the goods brought back—but that a positive result has not been obtained; yet, that result, though negative, is neither un-

important nor valueless; for, if the main fact has not been determined, our view has been cleared, and we have gained much knowledge respecting the intercourse between the nations of the east and the west."

I know of no better introduction to the following pages than these great words of Carl Ritter. I believe he has erred only, when he means, that the entire cycle of possibilities for the explanation of this historical tradition has already been exhausted in his own times. It is my intention to show in the following pages, that, hitherto, all attempts—Ritter's included—at solving the problem, have left the only possibility unheeded, which, I am convinced, can give a thoroughly satisfactory solution.

* * *

**Biblical Quotations.**

In order to enable my readers to grasp the question firmly, I will now quote all

passages in the Old Testament, in which the name, "Ophir," is mentioned. I will generally use the authorised version, but in essentials I will go back to the original text, and to the versions of the polyglot Bible, whenever the clear comprehension of the subject requires it.

*Genesis* x, 20-25: "And unto Ebor were born two sons; the name of one was Peleg, for in his days the earth was divided. His brother's name was Joktam. And Joktam begat Almodad and Sheleph, and Jerah, and Hadoram, and Usel, and Diktah, and Obal, and Abimael, and Sheba, and Ophir, and Havillah, and Jobab; all these were the sons of Joktan. And their dwelling was from Mesha, as thou goest into Sephar, a mount of the East."

The name of Ophir here designates a tribe of Arabs. Mesha is the modern Musa;

Sephar, later known as Dhafar, Dhofar, in the incense country, now the "Isphor" of the indigenes. The mount of the East is the high range of mountains called Faguer in the Ekheli language (see Ritter's Geography, vol. xiv, fol. 372). In this passage we have to do with a geographical indication, pointing towards the east of the Hadramaut (the region along the southern coast of Arabia). There is no evidence of its connection with the later accounts of Solomon's Ophir. All the passages of the Old Testament, following hereunder, refer to these later accounts.

I *Kings* ix, 27-28: "And Hiram sent in the navy his servants, shipmen, that had knowledge of the sea, and the servants of Solomon. And they came to Ophir, and fetched thence gold, four hundred and twenty talents, and brought it to King Solomon."

The Hebrew text spells the land referred

to in this passage "אוֹפִיר" whilst in the passage from *Genesis*, given above, it reads "אוֹפִר" In the Septuaginta the land in *Kings* is translated "Σωφειρά," whilst in *Genesis* it is written Οὐφείρ. The "talent" in the authorised version stands for the Hebrew Kikkar in the original text. Soetbeer calculates the Kikkar of gold at 42·6 kilograms, of the appromimate value of £5,600.

1 *Kings* x, 10-11: "And she (the Queen of Sheba) gave the King an hundred and twenty talents of gold, and of spices very great store, and precious stones; there came no such abundance of spices than which the Queen of Sheba gave to King Solomon. And the navy also of Hiram, that brought gold from Ophir, brought in from Ophir great plenty of algum trees and precious stones." (Algum—algumim, sandalwood or ebony?)

II *Chronicles* viii, 17-18. An account of the same events, evidently derived from the same sources, is given as follows: "Then (after the building of the temple) went Solomon to Eziongeber, and to Eloth, at the seaside in the land of Edom. And Huram sent him by the hands of his servants ships, and servants that had knowledge of the sea; and they went with the servants of Solomon to Ophir, and took thence four hundred and fifty talents of gold, and brought them to King Solomon."

Keil explains the difference in the number of talents in *Kings* and *Chronicles* by a clerical error, caused through the similarity of the Hebrew signs for 420 and 450.

II *Chronicles* ix, 9-10: "And she (the Queen of Sheba) gave the King an hundred talents of gold, and of spices great abundance, and precious stones; neither were there any such spices as the Queen of Sheba gave King Solomon. And the servants also of Huram, and the servants of Solomon, which brought gold from Ophir, brought algum trees and precious stones."

H. Ewald, in his "New Observations on the Navigation to the Gold-Land Ophir," in the report of the Royal Society of Sciences, University of Goettingen, No. 18, 1874, writes, concerning these accounts in *Kings* and *Chronicles* as follows: "We possess them only in abbreviated and disconnected extracts, but in the present state of our science we can certainly recognise, that these extracts have been made by expert historians from the State Annals, which were compiled soon after Solomon's death. These State Annals, as we now know them by these and many other extracts, and can take them at their real value, contained docu-

mentary reports, told in a simple manner, and free from all flattery of the kings."

The divergences occurring in the scriptural accounts given above find their explanation in the hundreds of years which intervened between the writing of *Kings* and *Chronicles*, the latter only from about the middle of the fourth century B.C. Therefore in the case of all such divergences we will adopt the accounts in *Kings*. In any case, however, we have to deal in all these accounts with firmly established historical facts.

A series of further passages may now be quoted as helping to render the subject intelligible.

I *Chronicles* xxix, 4: "I have given," says *David*, "three thousand talents of gold, of the gold of Ophir, and seven thousand talents of refined silver, to overlay the walls of the house withal."

*Job* xxviii, 12-16: "But where shall wisdom be found? and where the place of understanding? Man knoweth not the price thereof, neither is it found in the land of the living. The depth says, it is not in me, and the sea saith it is not with me. It cannot be gotten for gold, neither shall silver be weighed for the price thereof. It cannot be valued with the gold of Ophir with the precious onyx or sapphire."

*Psalm* xlv, 9: "Kings' daughters are among thy honourable women; upon thy right hand did stand the queen in gold of Ophir."

*Isaiah* xiii, 12: "I will make a man more precious than fine gold; even a man, than the golden wedge of Ophir."

*Job* xxii, 24: "Then shalt thou lay up gold as dust, and the gold of Ophir as the stones of the brooks."

In order to consider these accounts without prejudice, we must first of all state, that their writers simply mean by Ophir a gold producing country, and *Job* xxii, 24, gives ground for the conclusion that it was a country with rivers containing auriferous sand. This must be kept firmly in view, for it eliminates at once quite a series of attempted interpretations.

These accounts further show, that Ophir as a gold producing country was known to the Jews long before the time of Solomon. David already mentions Ophir, and he spent the enormous sum of 3,000 kikkars of Ophir gold, or, approximately, £16,700,000 on the building of the Temple. Finally, above all, Ophir occurs in these accounts as a well-known name, with which the readers are perfectly conversant. Not one of the writers deems it necessary to explain the word in

any way whatever. The knowledge of its meaning is everywhere presupposed. This circumstance is likewise of importance in our research, for it proves that quite a lively intercourse must have been carried on between the semitic nations of western Asia and Ophir, wherever it was situate, of which even the common people had knowledge.

We may therefore conclude, that under Ophir no distant region on the confines of the globe, as then known, was meant, but a country, which lay in the centre of the World's traffic. The name of Ophir was used as in our days we use the name of America. I observe, that, hitherto, this point has not been sufficiently considered in the treatment of the question. Because Ophir has become a myth to us, it was thought, that even in ancient times it lay in nebulous remoteness, whereas the indefinite manner, in which it is mentioned, proves, that it was in full vision of the semitic world as it then existed.

Only two passages in the Old Testament appear opposed to this view. They have played an important part in the Ophir literature, although the name of Ophir is not even mentioned in them.

In I *Kings* x, 21-22, it is written: "And all King Solomon's drinking vessels, and all the vessels of the house of the forest of Lebanon were of pure gold; none were of silver, silver was nothing accounted of in the days of Solomon. For the King had at sea a navy of Hiram; once in three years came the navy of Tarshish, bringing gold and silver, ivory and apes, and peacocks.

II *Chronicles* gives the following rendering of the same incidents: ix, 20-1, "And all the drinking vessels of King Solomon were of gold, and all the vessels of the house of the forest of Lebanon were of pure gold ; none were of silver, it was not anything accounted of in the days of Solomon. For the king's ships went to Tarshish with the servants of Huram ; every three years once came the ships of Tarshish bringing gold and silver, ivory and apes, and peacocks."

We observe, that the author of *Chronicles* has remoulded the account in *Kings*, in order to explain the idea "Tarsisship," and reports, that these ships sailed to a country or place named Tarsis. I will presently recur to this, and now only remark, that principles of criticism direct us to follow in such cases the older and simpler texts, and to reject the attempted explanations of later centuries. We will, therefore, take it for granted, that the account in *Kings* treats of the voyages of Tarsis ships, and not of

voyages to Tarsis. Gesenius and others annotate the text as follows: "Where Ophir is not named, but certainly meant." They do not prove it, but it is in itself most probable. As, however, it gives rise to quite a number of conclusions, it is convenient to ascertain at once, what is related and what is not related.

The passage in *Kings* reports only, that Solomon's fleet returned once in three years to its port of departure. Its meaning, however, is not, that the objective necessitated a voyage lasting one and one-half years. Is it not possible, that the object of the voyage required, that the ships remained a long time to fulfil their mission, *e.g.*, to collect quantities of gold by trading or digging? Or is it not conceivable, that they had to call in several countries on the shores of the Indian ocean to collect their miscel-

laneous return freight? Both hypotheses are well founded, and they destroy the demonstrative force of the account for the very things they were believed to prove. Further on we shall see, that Lassen uses the accounts for showing that Ophir was situate in India. I shall therefore have to refer to it more fully later on. Here I will, firstly, maintain, that it is not stated, that the ships required three years for the voyage to and from Ophir, and, secondly, that it is nowhere stated, that all the goods mentioned in the accounts were of Ophirian origin. This must be kept in view for testing their demonstrative force.

What was understood by Tarsis is not settled to the present day, and I have not been able to form a hypothesis founded on facts. In *Genesis* x, Tarsis is mentioned as the child of Javan, the son of Japhet, and

in history it occurs as the name of quite a number of Phœnician settlements: in Cilicia, the ancient name of Tunis, in Spain (Tartessos in the Greek tongue), and in Southern Arabia, as Promontorium Tarsis on the coast of Oman, along which, coming from the Indus, Nearchos sailed with Alexander's fleet. Quatremère, Tuch and Carl Ritter assumed, that Tarsis or Tarshish had the meaning of our "Thule," and signified extreme remoteness, and that Tarsis-ship stood for a large sea-going ship, corresponding with our "liner." This interpretation had already been adopted in the Septuaginta, where Tarsis-ship is translated "πλοῖον θαλάσσης" ocean vessel, a translation, which Luther has likewise adopted in his translation of the Bible. Failing a more precise explanation, I feel inclined to accept this interpretation as the basis for our further enquiries, and

will now quote the passages in the Old Testament, which name the starting point of the voyages to Ophir.

I *Kings* xxii, 47-50: "There was then no king in Edom, a deputy was king. Jehoshaphat made ships of Tarshish to go to Ophir for gold, but they went not; for the ships were broken at Ezion-geber. Then said Ahaziah, the son of Ahab, unto Jehoshaphat, Let my servants go with thy servants in the ships. But Jehoshaphat would not."

II *Chronicles* xx, 35-37: "And after this did Jehoshaphat, King of Judah, join himself with Ahaziah, King of Israel, who did very wickedly. And he joined himself with him to make ships to go to Tarshish; and they made the ships in Ezion-geber. Then Eliezer, the son of Dodavah of Maresha, prophesied against Jehoshaphat, saying, Be-

cause thou hast joined thyself with Ahaziah, the Lord has broken thy works. And the ships were broken, that they were not able to go to Tarshish."

We observe in these passages the same divergence in the question of Tarsis, which we observed above, and we adopt the same view regarding it.

The other divergence, consisting in this, that in *Kings* Jehoshaphat declines the alliance with Ahaziah, whilst in *Chronicles* he accepts it, and thereby wrecks his enterprise, is evidently the later addition of a priestly annotator. Our interest in these passages, the starting point of the voyages to Ophir, is clearly defined therein. It is unanimously recognised as Ezion-geber, on the northern termination of the Gulf of Akaba, not far from Eloth or Ailat, the ruins of which were discovered in 1829

by E. Rueppell. The port was within Solomon's jurisdiction, and was also used by his ally, Hiram. The voyage was thence made through the Red Sea, and it clearly follows that its objective, Ophir, must be sought for on the shores of the Red Sea, or on those of the Indian Ocean. Where Ophir was really situated is the great question, which has not been satisfactorily answered to the present day.

* * *

**Former views on the situation of Ophir.**

I will not enter on a detailed description of the fads and fancies, which the attempts at solving the Ophirian enigma have brought forth in the course of centuries. Of the interesting attempts I will only say that Calme (Dissertation sur le pays d'Ophir, in the Traites Geographiques, à la Hays, fol. 287) places Ophir in Armenia ; A. J. von Hardt (Diss. de Ophir, Helmstedt, 1746) in

Phrygia; Alderman (Diss. de regione Ophir, Helmsted, 1718) in Spain; Arias Montanus, Wilhelmus Postellus and others, in Peru; M. Lipenius (Martini Lipinii, diss. de navigatione Salomonis Orphirica, in Ugolini thesauro, vol v, fols. 343-87) relegates it with Flavius Josephus to the Malayan Peninsula (Aurea Chersonesus); Hadrian Rolandus (Dissertatio iv de Ophir, in diss. Misc. Trajecti ad Rhenum, 1706, fol. 186) and William Ouseley (Trav. London 14, vol i, fol. 47) to Ceylon; Macdonald (Asiat. Researches, vol. i, No. 17) to Sumatra. And Christopher Columbus was firmly convinced that he had found Solomon's Ophir in the West Indies. Reporting to the King of Spain on his third voyage, he writes: "The mountain Sopora (the name for Ophir, which, in the Septuaginta is written Sophora), which it took King Solomon's ships three years to reach on the island of Haiti, has now come with all its treasures in the possession of their Spanish Majesties."

Out of this chaos of conjectures and fables have risen in the course of centuries four attempts at solving the problem, which are represented by keen and ingenious scholars, and can claim their full value as solidly-established hypotheses. We must confront them with determination if we would hope to arrive at a solution of the Ophir question, that shall give satisfaction in every respect. An examination of the concrete reasons for each one of these four hypotheses will not help us towards the final solution of the problem, for all these attempts, as Gesenius and Ritter aptly remark, have only resulted in a larger or smaller measure of probability for one or the other theory. But their consideration will cast a bright light on our research.

**The Thule Hypothesis.**

The first theory in explanation of the meaning of Ophir claiming our consideration takes it as a collective designation of remote southern countries generally, analogous to the interpretation of Tarsis by Tuch and Ritter. According to this theory we would have to understand by Ophir something like Thule, and, consequently, the scriptural passages quoted above could not be applied to one definite country. This theory was upheld, *i.a.*, by Pater Joseph Acosta, Heeren, Tychsen and Zeune. The reader will observe the difficulty, for, whilst it gives the name of Ophir such a very wide meaning, it does not answer the question, whence Solomon and the Phœnicians fetched their gold. The assertion that Ophir means simply "remoteness" or

"South" does not help us, for the ships doubtless sailed to a definite country, and the real problem is not even touched on, with which remote or southern region the commercial relations were cultivated, which are so clearly indicated in the tradition of the Ophir voyages.

Therefore, we can pass over this theory, for the question does not so much touch the etymological interpretation of the word "Ophir" as it does the unravelling of the mystery of the Phœnico-Judæan sea voyages. Also, the unprejudiced reader of the scriptural passages quoted above will scarcely be able to reject the conviction, that the interpretation "south" does not do justice to the case, as will be seen when remoteness or south are substituted for Ophir in the texts. Besides, these indefinite terms appear untenable, when we consider the

definite indication of Ezion-geber as the port of departure. Should not the authors, who knew so accurately whence the fleet sailed, have known its destination more accurately than the ideas "remoteness" and "south" express it?

I therefore reject this theory, for two reasons. Firstly, because it does not go far enough, for, even if it were correct, the investigation must go beyond it to exhaust the historical side of the question; and secondly, it does not do justice to the original tradition, which, without doubt, meant, by Ophir, a distinct country. It is another question, whether Ophir did not originally designate a certain quarter of the globe which afterwards became the name of a certain country or continent. I will, later on, recur to this. Our first consideration must be, which definite country was meant

in the tradition as delivered to us in the Old Testament, and that point has not even been touched on in this theory.

**Arabian Hypothesis.**

In that respect it stands behind the other three theories. One of these, adopted by men like Edrisi, Abulfeda, Bochart, Niebuhr, Gesenius, Vincent, Gosselin, Volney, Seetzen, Rosenmueller, Keil, and more recently by Soetbeer, looks for Ophir in southern Arabia. There are, indeed, good reasons for it. We know, that on the coast of Oman a number of early Phœnician settlements have been traced. The Gulf of Persia, in ancient times, was a great field of Semitic commercial activity; on its shores was the cradle of the Phœnicians. Not only Strabo and Herodotus state this, but *Ezekiel* xxvii, 15, confirms it indirectly, when he relates that the Dedan were merchants of the City of Tyre. It has been

proved, and accepted by Ritter, that the Dedan were settlers on the Gulf of Persia. On that Gulf were still, in historical times, the commercial ports, Tyros and Arad, where Hiram obtained ships for Solomon, which were taken to Ezion-geber. In order to maintain, that these ships were taken from Tyre on the Mediterranean to the Red Sea, the monstrous improbability of their transport by land over the isthmus of Suez would have to be assumed.

In south Arabia was the residence of the Queen of Sheba or Saba, whose relations with Solomon are well known. The promontory of Tarsis, which was passed by Nearchos, is on the coast of Karaman. Further, Edrisi reports that two days inland from the ancient Szohar (the modern Sur), was an Ophir, which he calls Ofar or Ofra (Jaubert's ed. of Edrisi I, fol. 152); an Afir

in El Ahsa (also called Ghafar); a mount Ophir in Barhein (fol. 147). Does it not call to mind the above-quoted passages in *Genesis* x, where the tribe Ophir is mentioned in Hadramaut, between Saba and Havilah? Is it not extremely probable that the Old Testament meant under its two Ophirs one and the same object?

We may admit, that the simplicity of this interpretation is rather captivating, especially because it solves at the same time and in a very easy way, the Tarsis problem as above related. Nevertheless, the reasons for it are not weighty enough to raise them above the value of a pretty hypothesis.

The triennial duration of the Ophir voyage, which has been pleaded against this theory, does certainly not influence me, for it is nowhere stated that three years were required for the accomplishment of the voy-

age. But how can we take Arabia for a gold-country, which the biblical Ophir is before all things? There is no gold in Arabia, and consequently I hold, that this second theory falls to the ground. For the vast quantities of gold, which the scriptural tradition mentions, which Soetbeer values at £46,000,000 and a modern author (E. P. Mathers, in Zambesia, England's Eldorado, fol. 43), though without any foundation, at even £900,000,000, refute the explanation, that an Arabian intermediate trade was meant. How could Solomon have paid for such enormous values? With the agricultural produce of his own country? No, nothing but mining on a very large scale could have been meant or barter with untutored aborigines. Now, Soetbeer, relying on a passage in Agatharchides in his description of the Erythrean Sea, written in the time of

Ptolemaeus VIII (117-107 B.C.), attempts to prove, that South Arabia was really rich in alluvial gold, and that it was that gold, which was fetched by Solomon's expeditions.

I am of opinion that, in view of the geologically determined fact of Arabia's poverty in the matter of gold, such a statement by an Alexandrian writer, which can be easily explained by an error, cannot be considered as of any weight whatever. The conclusions, which Soetbeer draws from it are nothing more than unfounded combinations, for he cannot adduce a single fact showing that the Phœnicians and Jews got their gold from Southern Arabia. No traces of their presence in that region can be brought forward. Consequently Soetbeer's ingenious deductions are mere conjectures, and therefore the South Arabian theory fails, inasmuch as Ophir *is* a gold-country and Arabia is *not.*

Also, the fact that there never were elephants in Arabia, whilst ivory was second only to gold in importance as an Ophir article of export, speaks against placing Ophir in South Arabia. These facts, in their actuality, appear to me of so much weight, that they cannot be overridden by mere similarity of names and philological arguments. For we must bear in mind, that Arabian Ophir and Edrisi's Ofra were not known in very ancient times, and later on we will see, that several alliterations of the name "Ophir" occur in regions, which more closely correspond to the attributes of Solomon's Ophir.

In such researches as ours it is, above all, necessary to keep the essence of tradition in the foreground, for similarity or identity of names are always explainable by accidental causes. The essence of the Ophir tradition, as we have seen, consists in this, that it is

described as a gold-country, and therefore I reject the South Arabian theory.

**Indian Hypothesis.**

The third Ophir theory enters the field with even more weighty pretensions, and we must admit, that it has maintained its superiority until the present day. According to it Ophir must be located in the East Indies, and it is upheld by brilliant champions of all times, as by the Septuaginta and the Alexandrian School, by Flavius Josephus, Bochart, A. W. von Schlegel, and above all by Christian Lassen (Indische Altherthums-kunde, Bonn 1843, vol. i) and Carl Ritter (Geography, vol. xiv.) The scientific reasons in favour of this theory have been furnished by Lassen and Ritter, whose arguments we will have to consider rather closely.

The close examination of Lassen's and Ritter's arguments shows, that the leading idea of their theory is based on the follow-

ing conclusion Lassen's: "If it can be shown, that all the goods which Hiram and Solomon brought from Ophir, and also their non-Hebrew names are of Indian origin, then it is quite unnecessary to examine again the numerous conjectures regarding the site of 'Ophir.'" He means thereby, that it is then clearly proved, that Ophir must be sought for in India.

It cannot be denied, that this conclusion contains a large measure of probability. But it cannot claim, that it is evidence.

It is quite possible, that Indian names were adopted in Ophir for Ophirian products, as, *e.g.*, Indian names have been adopted in the East African Kiswahili. Or it is possible, that in very early ages Indian words became naturalised in the Hebrew language, perhaps in the times, when the Phœnicians still lived on the shores of the

Persian Gulf, or through intercourse with Arabian tribes.

It does not follow, that, because some of these things are called by names of Indian origin, Ophir itself must have been situate in India. But let us pass this objection over for the present. Lassen and Ritter themselves admit, that their argument does not possess the unconditional force of evidence, they only claim for it a high degree of probability.

But how about the names of the goods that were obtained in these Ophir voyages? They are enumerated in the passage quoted above from I *Kings* x, 22: "For the king had at sea a navy of Hiram; once in three years came the navy of Tarshish, bringing gold and silver, ivory and apes and peacocks." (Tukkhiim in the Hebrew text, peacocks or guineafowls?) and in II *Chroni-*

*cles* ix, 21: "For the king's ships went to Tarshish with the servants of Hiram; every three years once came the ships of Tarshish bringing gold, and silver, ivory and apes and peacocks." (Tukkhiim.)

I have already pointed out, that the force of demonstration of these two passages is weakened by reason of their liability to an ambiguous interpretation. Ophir is not mentioned in either of them, and opinions differ as to what has to be understood by Tarsis. It therefore follows, that the only thing left to us is to strengthen the probability of one or the other theory. But is there no doubt left regarding the assertion of Lassen, that all the names mentioned in these passages, which are not originally Hebrew, are derived from Sanskrit?

The biblical text has amongst such non-Hebrew roots "Koph" or "Kuph," ape.

This word Lassen, and with him Schlegel and Gesenius, derived from the Sanskrit "Kapi." This is, unquestionably, quite possible. But it appears to me much more probable, that the Hebrew "Koph" is derived from the Egyptian word "Kaph" for ape. The Jews probably made the acquaintance of the name and the genus "Ape" in Egypt. Here two derivations are possible, and the least that can be said about the matter is, "Non liquet," it is not clear.

It is exactly so in the case of the second root. Ivory is called in the Hebrew text "Shen Habbim," meaning "tooth of elephants."

Lassen derives Habbim from the Sanskrit "Ibha" elephant. But in Egyptian the elephant was called Ebu. What could be more probable, than that the Jews took their word from the Egyptian? It is not even

decided, whether the Sanskrit Ibha is not borrowed from the Egyptian. Be that as it may, if the Jews brought that word from Egypt into Asia Minor, it follows with absolute clearness, that it stood in no connection with the Ophir voyages. But that is exactly what Lassen concludes from it.

But the peacocks, the "Tukkhiim" at least prove Indian origin? The peacock is an Indian bird, whose habitat is not in Arabia or Africa, and its name in Sanskrit is "Cikki." From this, it is supposed, the Hebrew "Tukkhi" has been derived, perhaps by a roundabout way through the Malaharian "Togei." Thus argue Lassen, Gesenius and Ritter. It is a pity, that also this derivation is not quite beyond doubt. In Latin Gallina Afra, the African hen, meant "guinea-fowl," called Tukka (see Ritter, fol. 419), which is still one of the

most favourite game birds in all East Africa. Was it not possibly the guinea-fowl, which the writers of *Kings* and *Chronicles* meant by their Tukkhi? The flesh of these birds is delicious, and if their plumage is less gorgeous than that of the peacock, it nevertheless is a very handsome bird. In any case, even the guinea-fowl must have been a rarity for Solomon and his court, well worth transport from a distance. We see at once, even in this case, the "Non liquet."

But where is then the force of demonstration of all the argument put forward by Lassen?

The case is still worse with the derivation of the last non-Hebrew name in the scriptural passages. Algumim or Almugim is said to be derived from the Sanskrit "Valgu," meaning "Sandalwood." This purely etymological deduction appears to

be somewhat forced. Sandalwood is a spice, and the Algumim are expressly enumerated as timber.

In II *Chronicles* ii, 8, Algumim from the Lebanon is ordered by Hiram for the building of the Temple. When grew Sandalwood trees on Lebanon, and when was such wood used for buildings? Cedars were undoubtedly meant here. Consequently "Algumim" did not exclusively signify Sandalwood. Hence it follows, that Lassen's arguments are not decisive, but admit the interpretation that "Algumim" means valuable wood generally. And consequently Luther's translation "Ebony" may be correct, and the word "Algumim" does not help in proving, that Ophir must be searched for in India. If the Ophir fleet had carried Sandalwood to Jerusalem the conclusion would be admissible, that its objective lay in India, although

Sandalwood grows also in Central Africa. But, as "Algumim"* also means Ebony and other valuable woods, we may here again maintain the "Non liquet."

In the matter even the learned arguments of Ritter do not support Lassen. His philological analyses are in no sense finally demonstrative. Briefly, Lassen has *not* done, what he believed he could do, viz., prove "that all the goods, which were brought for Hiram and Solomon, and their non-Hebrew names, were of Indian origin." His arguments are refuted by the possibility of two distinct translations of the enumerated objects, viz., "Gold, silver, ivory, apes, peacocks, sandalwood and precious stones;" or "Gold, silver, ivory, apes, guineafowls, ebony and precious stones." The first translation

* NOTE.—In Algumim *al* is article, *im* form of plural; so there remains the root *gum*, which perhaps is simply identical with our word gum.

would admit the Indian, the second the African objective of the Ophir voyage.

For further arguments in favour of the Indian theory we must chiefly look to Ritter, who adduces a number of facts, the importance of which cannot be denied. We know, that the Septuaginta translates Ophir by Sophir and similar modifications; and Sophir is, according to the showing of the native Coptic lexicographers, the name for "India with its islands." Ritter however adds, "but H. Roland already observes, that this name, mentioned also by Hesychius as the Indian gold-country, is not of early Egyptian origin, but can only have been taken by the Copts from the comparatively modern traditions of the Alexandrian interpreters." (fol. 381.) This proves the fact, that Ophir was placed in India in the period of the Alexandrian School, a circumstance worth remembering.

Gesenius states "the Arabian translator of the Polyglot Bible has also rendered the Greek of the Septuaginta in *Isaiah* xiii, 11, by 'El Hend,' *i.e.*, India."

The Arabian translator of the historical books seems to translate more definitely the Syrian word "Ophir," in 1 *Kings*, 28, by "Dahlak," "which belongs to India" (Ritter, fol. 381). This Dahlak is said to be the modern Socotra.

Flavius Josephus has also adopted the identification of the Ophir in *Kings* and *Chronicles* with the Sophir of the Septuaginta and he maintains in his Archæology, that the Solomonic voyages had India as an objective, which was named "Sophira" in ancient times (Antiq. Judæorum viii, 6 4). This statement is very interesting, for it is remarkable, that even Hebrew science, which should have been able to form a sound

judgment of the past history of its own nation, points with such certainty to Ophir in India.

Bochart placed Ophir in the island of Taprobane or Ceylon ; H. Reland in the ancient emporium Upara, the modern Goa. Ritter says (fol. 384) : "This supposition is strongly supported by the circumnavigation of the so-called Arrianus, who describes the most important and flourishing intercourse in his time between the Red Sea and the western coast of India. He mentions, after Barygaza, which was then the chief objective of all India-bound vessels, the same *Οὔππαρρα* (Greek for Ophir) as the emporium next to it in importance and distance, and which shared the world's commerce with it (Arrianni Periplus, mare Erythi, ed. Hudson, fol. 30).

"For the very early commercial inter-

course between India and anterior Asia, whose indigenes (Arabs and Persians) are called in Sanskrit, "Javanas," speak also the Sanskrit names of certain commercial commodities (*e.g.*, Javaneshta, tin; Javanapriya, black pepper; Javana, incense).

"In these monuments of primitive commercial relations, even ancient myths of the Egyptian Sesostris and the Indian Bharathas have been indicated, and they appear to have been fused into each other in the very earliest times" (Ritter, Geography v, fol. 442.)

All this supports materially and historically the theory, that Ophir was situate in India, although it does not absolutely prove it. For, why could there not have existed at the same time a very early intercourse with India, and another with an Ophir situated elsewhere? Ritter, with

great historical intuition, indirectly admits the possibility: "Another necessity for our geographical aim consists in this, that we not only become conscious of the real position of Ophir in India, and of the special facts in connection with the Solomonic voyages, but also of the entire influence of such remarkable events, yea, even of the entire group of similar voyages and commercial intercourses between the nations, in the centre of which, between east and west, this Ophir is placed, or which, chronologically, is the first known to history."

Lassen and Ritter recognise the name Ophir in the people called Abhira on the mouth of the Indus. "It was the nearest Indian coast for the Phœnicians, and here they could find stores of the goods of the north and the Himalayah, such as gold and bdellium, and those of the south, as sandal-

wood, &c." Even in our days, a tribe in that region is called "Ahir," a word which originally signified "cowherd."

"These Abhira, with many others of their tribes, which at the same time represented different castes, were also settled in the northern Punjab, but migrated thence in a southern direction, when Brahmanic colonies of the very earliest times still led a peaceful pastoral existence, and there these Abhira first reached the coast." "They must therefore belong to the oldest Brahmanic tribes, descendants of Indian Aria, and held sway in Solomon's time over that coast, so that the name Ophir must have been derived from them. Their migration to the mouths of the Indus must have taken place within the second thousand years B.C." (Ritter's Geography, 391 and 392.)

Then Ritter proceeds with brilliant in-

genuity, and with his vast knowledge, which embraces the entire world, to furnish the proof, that all the goods enumerated in the Ophir voyages were procurable at the mouths of the Indus. It is not necessary to our research to follow him in the details of his arguments. It is sufficient to state, that Carl Ritter has proved, that gold, silver, ivory, sandalwood, peacocks and precious stones could be procured in the country of the ancient Abir, and that, consequently, it could well have been the objective of the Ophir voyages. Nevertheless, the question is not solved thereby, how the Jews could have paid the Abir for such enormous quantities of gold as those mentioned in the Old Testament.

Ritter himself cannot explain how an agricultural country like Solomon's Israel could have paid a pastoral people by barter

for such vast treasures. "Nothing is said of the outgoing freight, with which these goods were bought or exchanged; we cannot, therefore, form a conjecture as to the requirements of the Ophirites. And this causes a certain onesidedness of the whole method of demonstration. The authors of *Kings* and *Chronicles* have not furnished a complete account of the commercial relations, therefore they have not even stated, at which prices the goods were obtained, and still less have they indicated the locality of the market" (398-9). I should like to add, that the authors do not even state, that the goods were got by commerce, but leave it an open question, whether the precious metals and stones were obtained by mining, the ivory by way of tribute, and the apes and Tukkhiim by capture. But therewith

Ritter's arguments lose much of their force of demonstration.

I am personally convinced, that mining enterprise was the main object of the Ophir voyage, if such vast stores of gold as 420 or 450 kikkars represented the result of a single expedition. My opinion is in some measure confirmed by a note in the fragment of Eupolemos (in Eusebius Præp, ev. ix, 30), which relates that King David sent miners to an island called Urphe, according to Gesenius more correctly called *Οὐφρή* or *Οὐφήρ*, on which were many gold mines, who brought gold thence to Judæa.

It will be conceded that this fact of mining operations must speak against the position of Ophir in India, for the Arian tribes, which in those times just there began their career of conquest, were not at all likely to have suffered any military occu-

pation for the protection of mining undertakings. If, therefore, we maintain, that the Ophir gold was got by mining, we must look for Ophir in regions inhabited by less valorous races. Those, however, were not to be encountered on the eastern, but on the western shores of the Indian Ocean.

That the conjecture of the acquisition of these masses of gold by barter is a difficult question for the Indian theory, is admitted by Ritter himself. As we have seen, he cannot explain the nature of the goods, with which the agicultural Jews could have traded in India, which is so rich in corn, and he is of opinion, that this point rather favours the African Ophir theory. He refers to an interesting note by Cosmas Indicopleustes (Cosmas Indicope. χριδανίκης τοπογραφίας Fragm. fols. 6 and 23 in M. Thevenot, Relations de divers voyages curieux, Paris, 1696),

who relates, that "the Hymyaritic traders of his time, who were under the protection of the Axomitic Kings in the country of the Agau, ('Αγαῦ), obtained by 'dumb' commerce for pieces of beef, iron and salt, which they placed on bushes, small ingots of gold from the dwellers on the coast of Zanzibar; that they returned home laden with gold after trading for thirty days without interpreters, and that the voyage there and back took six months." That is a picture of real life in East Africa, which applies even to present times. Ritter adds: "Such goods would not have been of any value in commerce with Indians, who do not eat meat, and to whom the cow is holy. Also, the Indian Steel (Wuz) was, on the authority of Etesias, better than any steel that could have been imported from the Occident." The attentive reader will agree with me, that this

difficulty, viz., the explanation of the getting of the Ophir products, casts a serious doubt on the Indian Ophir theory generally. Ritter himself could not surmount it. If we would finally characterise the arguments of Lassen and Ritter, we could claim to say, that their deduction from the philological point of view, on which they chiefly rely, appears unsuccessful ; but that the other material, with which they support the question, especially on the authority of the Indæo-Alexandrine school, strongly bears out the Indian theory. They have not conclusively proved their views, but we must admit, that, even in view of Theodore Bent's latest discoveries in south-east Africa, the Indian stands as a very real alternative possibility by the side of the east African hypothesis.

**East African Hypothesis.**

The latter places Ophir beyond Sofala in the regions of the Zambesi. Centuries ago

it has found eminent defenders, especially in Dapper, Th. Lopez, J. Bruce, Montesquieu, D'Anville, Robertson, Schultes, Quatremère, Krapf, Carl Mauch, Petermann and others. The Portuguese conquerors in particular were convinced, that they had Solomon's Ophir in those regions. But, remembering, that Columbus believed, he had found it in Hayti, we must not attach too much weight to that fact.

The Sofala theory had lost its credit entirely with the more eminent scholars of the first half of this century, after exciting the imagination of several centuries. Gesenius says, in his brilliant essay on Ophir (Allg. Encyclopædie by Erch and Gruber, 3rd Section iv, fol. 201): "Apart from these fancies the choice is left us only between a province of Southern Arabia and India, for Bochart's opinion" (Phaleg, ii, fol. 27, com-

pare Michaelis, Spicileg, ii, fol. 185) "which assumes two Ophirs, one in Arabia the other in India, and the supposition of Grotius and Huëtius, re-established by Bruce, according to which Ophir was situate on the coast of East Africa, the Arabian Sofala, the modern Zanzibar, and Mozambique with its gold-district Fura, have far too little probability in their favour ; in the last case for this reason alone, that its gold mines are nearly two hundred Spanish leagues inland." (See Hust, Commentaire sur les Navigations de Solomon in Traites geographiques, ii, fol. 65 ; D'Anville in Memoires de l'Academie, xxx, fol. 83 ; Bruce's Voyage to Abyssinia, i, fol. 479 of the German translation ; Schultes, Paradise, fol. 86, 296, 309. On the other side, Salt, Voyage to Abyssinia, fol. 102. Salt has visited the country and formed his counter-arguments especially from the nature of the locality. Vincent, Nearchos, ii, fol. 352; Tychen, zu Bruce's Reise, vol. v, fol. 329 of the Göttingen edition).

Ritter likewise attacks the views of Bruce thus: "The supposition of a Sabæan dominion extending as far as Sofala is quite unfounded, in so far as Bruce, adopting a myth of the indigenes, ascribes the ruins found, as Dos Santos relates, by the Portuguese invaders near the Tete gold mines, to the queen of Sheba. They are said to be covered by inscriptions in unknown characters. and bear the name of 'Fura'; and therefore it is Afura, and Solomon's Ophir in Sofala. Bruce goes even further, when he compares it with Simbaöe in the Empire of Monomotapa (Simbaöe meaning residence, which is said to correspond with the 'Agisymba' in Ptolemy's Aethiopia interior, lib. iv, 9, fol. 115.) But such gold mines

and remains of ancient buildings must with much more reason be ascribed to a much more recent Arabian commerce for gold, until testimony of a widely different nature has been adduced in favour of their Solomonic origin."

The testimony of a "widely different nature" has, as we now know, been quite recently brought forward. The rediscovery of the ruins of Simbaöe, or Zimbabye, by Carl Mauch, on September 5th, 1871 (see additions to Petermann's Mittheilungen, April, 1874) and their scientific examination by Theodore Bent (The Ruined Cities of Mashonaland, London, 1892) have furnished the proof, that in these ruins in the Zambesi regions we have to do with buildings, that date back at least to the times of Solomon, and are of Phœnician origin. These buildings represent temples and fortresses, and

were evidently erected in connection with gold-mining enterprises. I will later on refer to this in a more detailed manner. Here I will only say, that these discoveries have established a very solid basis for the African Ophir theory. If Gesenius and Ritter had known the result of the most recent investigations, they would unquestionably have abandoned their attitude of opposition against this theory, and conceded to it at least the same claim to attention as to the Arabian and Indian theories.

Nevertheless, these latest finds in Zambesia do not finally solve the Ophir question. They show, that ancient Phœnician-Sabæan mining operations were carried on there, of which the preserved ruins may be taken as documentary evidence; but that these mining operations were carried on in connection with the Hiram-Solomonic voyages, they

only make probable; they cannot conclusively prove it.

Bent, the best expert in the matter of the Mashonaland ruins, does not dare to consider them as the actual solution of the Ophir question. He says: "Regarding the doubtful question of Ophir, I do not feel it necessary to enter into its *pros* and *cons*. Mashonaland may, or may not, have been *the* land of the Ophir voyages; Ophir and Punt may, or may not, be the same thing; and both may have been there or elsewhere. As far as I can see, there is not sufficient evidence to establish a theory on these points, which could satisfy the more critical methods of investigation, which in these days are applied to subjects of this nature.

"All, that we can firmly maintain, is, that there the ancient Arabs obtained great quantities of gold. But, as gold was of common

use in prehistoric times and was largely used many centuries before our era, no doubt can be entertained, that its production must have been enormous and carried on in more than one place." (Bent, as above.)

This observation hits the nail exactly on the head. It measures accurately the importance of the influence, which the latest discoveries in South Africa exercise on the question of Ophir. They strengthen the possibility of finding in South-east Africa the objective of the Solomonic Ophir voyages, and invest this theory with a high degree of probability. They have raised the East African Ophir hypothesis to about the same level, on which the Indian theory was placed by Lassen and Ritter. It is possible, that Ophir was situate in India, but it can be that it must be looked for in East Africa. The existence of ancient Phœnician commercial

relations with both regions are now proved. But the material, that has been collected, does not go any further.

The question now arises, whether we cannot approach the subject from quite a different and perfectly new point of view, from which we can procure direct evidence. All these considerations and deliberations do not reach the kernel of the question, but are only of the nature of circumstantial evidence. I do not believe, that any conscientious jury in the world would give its verdict of "Yea" or "Nay" on the basis of the matter, that has thus far been brought forward. Is it not possible to force the ancient word Ophir itself to make a confession of its identity, and therewith finally to answer the question? I believe there is such a possibility, and through it I approach the crux of the entire investigations.

* * *

**Ophir identical with Africa.**

In a little work which I have quite recently published, entitled, "Equatorial and South Africa after a map of 1719" (fol. 10), I have raised the proposition, that our name of Africa (A-F-Rica) contains the ancient root of Ophir (Aleph, Phi, Resh), and that therein the key must be sought, by which the Ophir mystery can be really revealed. My further investigations in this direction have strengthened this conjecture in every direction.

Gesenius (Handwoerterbuch uber das alte Testament, fol. 20a) has adopted Sprenger's interpretation of Ophir as identical with the Arabian Afir (South Arabian Ofer), meaning "Red." The root of the Latin word Africa is Afer. Afer is the original name for African. (See Cicero ad Qu. Fr. i, 1, 9, 27; Sall. Jug. 18, 3; Liv. 29, 3, 13; Eutr. 2, 19.) Afri are the Africans.

Derivations from this root are the adjective Africus and Africanus; from the first is derived Africa, originally "Terra Africa." (Afirian country. See George's Lat. Dict.) The unprejudiced reader will concede that the sequence: Ophir, Afir, Afer, Africus, Terra Africa, and Africa is more than ordinarily suggestive, and points straightly towards the elucidation of our problem. It must be observed that primarily the Latin name "Afer" for African was applied to the inhabitants of the Phœnico-Carthaginian Province, and was only in a later period extended to the entire continent. We may be sure not to err, when we assume, that the Romans adopted this name during their earliest relations with Phœnicians and Carthaginians.

In order to form an accurate decision on this point, the following has to be made quite

clear. Hitherto there existed no philological derivations of the name "Africa." Its meaning and derivation have not been known to anybody until now.

It is said in "Nouvelle Géographie Universelle, x, 1, 2," as follows: "Quant à ce mot d'Afrique, appliqué maintenant a l'ensemble du corps continental, on ignore quelle en fut l'origine." (*i.e.*, As to the name Africa, which is now applied to the entire continent, its origin is unknown).

Well, I submit that my derivation of the name Africa from the ancient Hebrew word Ophir *is* the derivation which has been sought for centuries. Further, until this day the learned men of all nations were of opinion, that neither Phœnicians nor Jews had a comprehensive name for that part of our globe, which we call Africa. Yet it is a fact, known all over the world, that the Phœnicians

visited Africa in the very earliest times, and that their commercial enterprises simultaneously embraced the north, the east and the west of that continent. Can it be seriously believed, that they had no comprehensive name for regions, the connection of the parts of which they must have perfectly understood? My theory, that Ophir was in the earliest times the Semitic name for Africa as a whole, solves this question at once.

The matter stands thus: On one side rises the name Ophir out of the mist of primitive times as signifying a region, which was then generally known, but the position of which we do now not know; on the other side there is the Continent of Africa, which was known in the earliest times to the Semitic tribes of north and south, and with which they entertained a lively commercial inter-

course, and yet we do not know, what name it bore in the Semitic world. It now appears, that the root of the ancient Semitic name "Ophir," the bearer of which we do not know, has been preserved to the present day in our word Africa, the ancient Semitic equivalent of which is lost. Does not the next step follow quite naturally, viz.: to find the bearer of the enigmatical name "Ophir" in Africa, and the name of nameless Africa in Ophir? And do we not thereby solve this primitive problem in an obvious and natural manner? With one effort two interesting enigmas affecting the history of mankind are solved, and their solution is so extraordinarily simple that, once in possession of the thread, we can only wonder that the most ingenious brains of the last twenty centuries were unable to solve them long ago. However, the old story of the egg of Columbus finds

its application to all the branches of human thought.

* * *

**Etymology of the word Ophir.**

Regarding the etymology of the root, which is the basis of the name Africa, I have already mentioned, that its meaning is "Red." In explanation of the word I could state, that the African coast opposite Arabia, as far as Cape Gardafui, is of an intensely red colour ; also, that red Laterite is the principal geological constituent of Central Africa, facts, which would sufficiently explain the name. Thus, Africa would mean "Redland," analogous to the name Albion, which its chalk cliffs have conferred on England.

But I believe, that we are in the position to give another and more profound interpretation.

Though we venture with it into the realms of probabilities and combinations, it

will give us opportunities for entering on several problems of the primitive history of mankind.

Gustav Schlegel, in his Uranographie Chinoise (Leiden, 1875), explains, that the Chinese distinguished the quarters of the globe by different colours: "La partie orientale fut nommée le domicile du dragon *bleu ;* la partie boréale fut nommée le domicile du guerrier (or de la tortue) *noir ;* la partie occidentale fut nommée le domicile du tigre *blanc ;* et la partie mériodionale fut nommée le domicile de l'oiseau *rouge*," (fol. 1).

In other words : the north was marked in *black ;* the east in *blue ;* the west in *white*, and the south in *red*, and in addition thereto each of these quarters had its hieroglyphic symbol in the form of an animal. This is the foundation of the entire system

of Chinese astronomy, which Schlegel conjectures originated 17,000 years B.C., and was brought by Chaldæans and Phœnicians to the Egyptians and Greeks. Schlegel undertakes to prove, that the sidereal maps of all nations, and the names of the constellations have their origin in China. He maintains, that the names of the constellations do not signify pictures of the objects, which they represent, but only hieroglyphic Chinese symbols for quite different ideas. These ideas represent the occupations of a primitive agricultural people, and the names of the constellations are really intended to form an agricultural almanack; that is to say, they represent those agricultural labours which were seasonable 17,000 years B.C. in Chinese latitudes in the signs of the respective constellations.

Everybody, who glances at the map of

the heavens, and at a clear nocturnal sky, knows that the constellations do not represent the objects, the names of which they bear. And this renders the assumption most probable, that they originally had symbolic or hieroglyphic significance.

I need not follow Schlegel any further in this matter. It may be, that Richthofen is right, who, in his admirable work on China, charges Schlegel with going too far in his deductions. For our particular purposes it is quite immaterial, whether Schlegel is right, when he says, that Chaldæans and Egyptians obtained their knowledge of astronomy from China; or Terrier de Lacouperie (Western Origin of the early Chinese civilisation, London, 1894), who, conversely, derives the origin of Chinese civilisation from the west; or, finally, von Richthofen, who explains the homogeneity of the ele-

mentary astronomical principles of the Chinese, Indians and Semites, by their common origin in the Pamirs. It suffices to know, that the same astronomical ideas were common to the three great races, Mongols, Arians and Semites ; that there was a time, when red meant south, and when the symbol or hieroglyph signifying south was the image of a "red bird."

Schlegel (i, fol. 69) writes as follows regarding this subject : "On the celestial map of the ancient Chinese the south is occupied by the picture of a gigantic bird, which spreads from Gemini to the Rowen, and is called the Red Bird. The *red bird*, or "*Foung*" is *China's phœnix.* Moreover, there are several phœnixes : the blue phœnix, the red phœnix, the yellow phœnix, and the rose-coloured phœnix. The blue bird is proper to the element Wood, and to

the east; the red bird (*the* phœnix) is proper to the element Fire, and to the south; the yellow bird to the element Metal, and to the west, and the black bird to the north and the element Water."

This is indeed interesting. If it be true, that these ideas were common to the races of eastern and anterior Asia, then they explain the names of a number of seas, interpretation of which has been a moot point for centuries.

Where has the Red Sea obtained its name, its water being nothing less than red? Why do we speak of a Black Sea in the north of Asia minor? Why did the Arabs call, and why *do* the Turks still call the archipelago to the west of Asia Minor (Bahr-en-send in Turkish) the White Sea?

The colour of the water of these three seas being the same, it becomes an obvious

necessity to look elsewhere for an explanation, and Schlegel's theory looks plausible enough for the purpose. It is true, the Greeks first named the Red Sea, *'Ερύθρα θαλάσσα*, and the names Black Sea and White Sea are of still more recent origin. The Hebrew name for Red Sea is "Yamsuph," which means the "reed" Sea. But, it is possible that the Greeks obtained their name from the Chaldæans, whilst the Jews coined their own special name. However that may be, the existence of these names is sufficiently noteworthy.

A second point must be considered. In ancient Egyptian representations the country Punt or Phoun is mentioned, and Sanch Kara of the 9th dynasty, about 2,500 B.C., and the Queen Hatsepsu, of the 18th dynasty, about 1,600 B.C., sent expeditions thither. They brought back gold, ivory,

several kinds of valuable timber, leopard skins, and two kinds of monkeys, among them baboons, too. The interpretation of the word Punt is just as much disputed as that of Ophir; but judging from the above-mentioned products, it is probable, that under Punt an African country has to be understood, most likely Somaliland (see Soetbeer as above, fol. 52). I lean towards the conjecture that the root Phoun is identical with the Latin Puni or Pœni, which, again, is the same as the root of *Φοῖνιξ* or Phœnicians. Now, the root of that name is *φοῖνος*, which means *ἔρυθρος* or red (see Pape Woerterbuch der griechischen Eigennamen). In ancient interpretations the Phœnicians were already called the Red or Reddish.

We have thus three facts which all converge in the same point: three names, the roots of which mean red, and all three point to the South: Afir or Ophir; Phoun or Phœnix; and 'Ερύθρα θαλάσσα. The second of these names has two meanings, a tribe, and the red bird, which symbolised the South in the Chinese celestial maps, and was there called Phoung. The Erythrean or Red Sea (see Pape as above, fol. 389B), formerly included that part of the Indian Ocean, which, south of Asia, reaches from the Arabian Gulf to the Island of Taprobane (Ceylon). It was only in much more recent times, that the name was restricted to the modern Red Sea. Would it be too daring to attempt an explanation of all these facts from one single point of view? Ophir or Africa is Southland; Punt or Phoun is likewise Southland; Puni or Phœnicians are the Southmen, the counterpart of the Normen, who entered the arena of history in much more recent times; the Erythrean,

or Red Sea, is the South Sea, the counterpart of our North Sea (the German Ocean).

We must here bear in mind, that the Phœnicians were originally settled on the Persian Gulf, and consequently migrated from the south to their home on the Mediterranean. Now, this bird Phœnix or Phoung of the Chinese is the symbol of the torrid South, and the root of all these names identifies South with Red. Perhaps the ancient myths of the Phœnix had their origin in misunderstood explanations on the ancient Chinese sidereal maps.

To us the most interesting result of understanding these different facts is this, that originally the words Punt and Ophir had a meaning which points, pictorially, to the idea "South," "Ophir or Afir" appears to be the ancient Arabian name ; Punt, Phoun, or Foung (Greek, φοῖνος), the Chino-Egyptian

name for the same idea. If we maintain this possibility, we can, perhaps, identify the tribe Ophir, in *Genesis* x, 10, who dwelt in southern Arabia, with the Puni or Phœnicians, and if the Egyptian Punt or Phoun refers to north-eastern Africa, we may think of the tribe of Afer or Afar, the modern Danakil, opposite Bab-el-Mandeb, who, according to their own traditions, crossed over from Arabia, and now form a sub-division of the Somali. Afar would then be Arab for the Egpytian Phoun. In the Noveau Dictionaire de Géographie Universelle (Paris, 1892, vol. v, fol. 41) it is said, after pointing out, that the colour of the Red Sea does not explain its name: "The majority of modern interpreters derive its etymology from the name of the red people, to which originally a vast number of the aborigines belonged. The Hebrew 'Edomites' Himyiar, derived

from Ahmar in Arabic (and, following Gesenius, I add Afir and Ophir), have that meaning; also Punt or Phoun, the name of an important Canaanite tribe." "These Canaanites migrated to the Mediterranean, where they became the Phœnicians of the Greeks, and the Puni or Pœni of the Romans." The latter circumstance recalls the fact, that to the Romans the Africans continued the Afri, which we explained by their close relations with the Carthaginians.

Therewith the various ancient problems: What is Ophir? What is the meaning of Punt? Whence is the name Red Sea derived? What is the significance of the ancient myths of the bird Phœnix? are conclusively proved to represent the same problem, and in all probability that problem finds its solution in its reduction to the ancient Chinese identity of Red and South,

and their hieroglyphic expression by the bird Phœnix on the Chinese maps of the heavens. If this be true, Africa would have had in ancient time the same meaning as Australia in ours, viz., Southland. I am quite conscious, that with this part of my exposition I cannot step beyond the limits of possibilities, but I trust, that my deductions have not gone beyond the scientific boundaries of permissible combinations. If my conclusions are correct, then they disclose a vista of the cultural relations of East Asia with Anterior Asia, and East and North Africa in the most primitive times of man's history, far beyond the limits of even our traditional lore.

* * *

**Result of our Research.**

At this point of my exposition it becomes necessary to understand accurately and definitely the real fundamental idea of my

mode of demonstration. Lassen, and with him Ritter, took the following stand-point : " If it can be shown, that all the goods which Solomon and Hiram imported from Ophir, and also their Hebrew names, are of Indian origin, then it is quite unnecessary to examine the various conjectures regarding the position of Ophir." On the other hand I contend : "If it can be proved, that our present word Africa is nothing else but the Latin adjective form of the ancient root of Ophir, then the question of the interpretation of this name in the Old Testament finds a final and satisfying answer." I believe I have proved, that neither Lassen nor Ritter have done justice to the task, which they had taken upon themselves ; and for my own part I hope, that I have succeeded in furnishing evidence, which proves my hypothesis. And, therefore, I claim the right

of maintaining, that the word Ophir in the Old Testament must be translated with our *own* word Africa.

If the reader will take the trouble to substitute this new idea in the passages of the Old Testament quoted above, he will find, that they will lose their strange and almost mysterious character, *e.g.*, 1 *Kings* x, 2: "And the navy also of Hiram, that brought gold from Africa, brought in from Ophir great store of Almug trees and precious stones"; and *Job* xxii, 24: "Then shalt thou lay up gold as dust, and the gold of Africa as the stones of the brooks." It is now made clear, why the ancient authors could everywhere use the word Ophir without explaining it, for the continent of Africa was scarcely less known to the Semites of Solomonic times than to the average European of our own times. It also renders it intelligible, how it has occurred,

that a name known in former times to everybody has to all appearance been totally lost. As a matter of fact it has not been lost, but it survives for all times in its Latin rendering. But the European nations, and above all the Romans, who supplanted the Semitic nations in the foreground of history, and continued the use of the primitive name "Ophir" in its transformation "Africa," were far too little conversant with Semitic literature to retain the knowledge of the derivation of that unintelligible word from its primitive root. And the Semites themselves, out of whose range of vision Africa had disappeared more or less during the revolution of the World's historical relations in the period from the downfall of Solomon's power to the battle of Zama, could afterwards not remember, where the Latinised name which they had meanwhile adopted had its real

origin. Many analogies of such a case could be adduced, *e.g.*, from the history of the development of the French and German languages. Our explanation, from every point of view, absolutely solves the enigma of the Ophir question.

**Objective of Ophir Voyages.**

As we now occupy this strong position towards this enigma, we shall be able to answer with greater critical certainty than before, which was the real objective of the Ophir voyages, especially those under Solomon's auspices.

For, when we know, that it must have been an African region, to which the vessels sailed, the decision is no longer a matter of difficulty. At three points in Africa, apart from its general name, names occur in historical times which phonetically remind us of the root Ophir, and in each case in regions, where it can be shown, that they stood in

direct relations with Phœnicians and Arabs. We will first examine those regions, which are connected with our question.

The first region is, as we have already observed, the district of Carthage, the inhabitants of which were called *Afri* by the Romans. The Romans obtained that name undoubtedly from the Puni, with whom they entertained very ancient commercial and political relations. It is quite out of the question, that this district could have been the objective of the Solomonic Ophir voyage, and I need not furthur discuss it.

The other region, which could enter into the question, is the district of the Danakil on the coast of Africa, opposite Bab-el-Mandeb and Yamen. The Nouvelle Géographie Universelle, x, p. 298, says of them : "Dans l'espace triangulaire entre la chaîne ethiopienne, la mer rouge et le cours

de l'Aouach les gros des habitants, nomades ou sédentaires, constitue la nation des Afar ou Afer, designés plus communement par les Européens sous le nom de Danakil." (In the triangle between the Ethiopian Chain and the course of the Hawash the majority of the inhabitants, nomadic or settled, belong to the nation of the Afar or Afer, generally called Danakil by Europeans.)

The belief prevailed, that in this district of the Afer or Afar the Punt or Phoun of the Egyptian Pyramids could be recognised. If the Ophir of the Solomonic epoch could be placed in this region, their long-disputed identity would be proved, but we have no real foundation for such a belief. Traces or monuments of ancient Phœnico-Sabæan mining enterprises are not preserved there. They unquestionably formed at one time part of the Sabæan dominion. But such

general assumptions have no longer a claim on serious consideration. In the present phase of the Ophir investigations we require quite a different, a positive, basis of fact and tradition, in order to build a new theory upon it.

**Zambesia.** In this respect the over-weighing force of actual facts, which recent explorations of the coasts and regions south of the Zambesi have procured for us, places the district of the ancient Afar quite in the background. Here we have such abundance of material that, as we are entitled to treat of Ophir as an African region, all serious doubts regarding its position are at once dispelled.

**Sofala.** First of all we still find here to-day the bay of Sofala or Sofara, as, according to Mauch (Ergaenz. zu Petermanns Mitteil, vol. viii, April, 1874), the name is still pronounced in the interior. As we have seen,

this name is only the Egyptised form of the Septuaginta for the Ophir of the Old Testament. The Septuaginta translates Ophir in different places with *Σουφίρ, Σουφείρ, Σωφίρ, Σωφείρ, Σωφηρά, Σωφαρά.* In Arabic L and R are interchangeable consonants, and thus originated the name Sofala. According to Gesenius this Græco-Egyptian form of Ophir originated with the prefix Sa, which in Egyptian meant land. Sa-Ophir changed into Sophir, with its Greek transformations, meaning literally "Ophirland," or the same as the Latin "Terra Africa."

To be sure, as Reland relates, the Indian Upara or Uppara and the Malabar Supora were also called Sofala up to the time of Abulfeda (1300 A.D.; Abulfeda Tab. XI, India in Buesching's Magazine iv, fol. 272) but with the adjective Indian. It necessarily follows thence, that the African was the

original Sofala, for otherwise it would have itself received a distinguishing adjective, perhaps "African."

It may have been, that sailors coming from the real Sofala, opposite Madagascar, gave, for unknown reasons, the name to the Indian locality.

There are several other alliterations of the name in the regions of the Zambesi. Between Limpopo and Zambesi flows the river Sabi or Sabia, which has borne its name since primitive times. About 225 kilometres from Tete is the big mountain Fura, the name of which the ancient Portuguese already recognised as a mutilation of Ophir. This mountain is very rich in gold. "To-day can still be seen ringwalls of the height of a man, one fitted into the other with marvellous art, without lime and the use of chisels." (See my work, Aequatorial

und Sued Africa nach einer Datstellung von 1719, fol. 11.) I would not place too much weight on these facts, if Sofala were not the export harbour of a primitive gold district κατ' ἐξοχήν, which to this day belongs to the largest gold-producing countries of our globe. As a matter of fact, until the discovery of Australia and California, the district between Limpopo and Zambesi was *the* gold country of our globe. "No country on earth," says J. Scott Keltie (the partition of Africa, London, 1893, fol. 9, 10) "known and accessible to the ancient nations, not even India nor Abyssinia, nor any country on the Mediterranean can be compared in their production of gold with the Zambesi regions." Here, in the middle ages, the gold country Motapa, a name which is preserved to the present age in the "Matoppo Hills" near Bulawayo, was located, where the Portuguese obtained their gold. Here are to-day the gold mines of Mashona and Manica-land.

It follows from this, firstly, and before all, that the essential quality of the biblical Ophir, its character as a land of gold, is completely met by the Sofala regions.

This is confirmed by an important circumstance : the fact, that in the Hinterland of this coast have been found the ruins of grand temples and fortifications, in connection with mining works, which are on the authority of their most recent and conscientious examination by Theodore Bent (The ruined Cities of Mashonaland, London, 1892), evidently of Phœnico-Sabæan origin, and of an age anterior to the times of Solomon. Here we have indeed, what we did not have in India : documentary evidence of a primitive and most extensive gold production, carried on by the nations, which are expressly mentioned in the Old Testament in connection with the Solomonic Ophir voyage, viz., the Sabæans and Phœnicians. How far-reaching this production of gold was, is shown by the calculation of F. W. Fairbridge, according to which these old workings extended over an area of about 4,000 English miles. It will be admitted, that these sober facts weigh very differently, when compared with the considerations, conjectures and possibilities, which could be pleaded for Arabia, India and other countries.

It is true, these ancient gold mines are not directly situated on the coast, but about 300 kilometres inland. But this cannot any longer be raised as an objection, after it has been proved, that we have not to do with one single mining expedition, but with the

organisation of a lasting occupation of the country for mining purposes protected by a chain of fortifications, that we have here a colonial enterprise of the first rank, raised above all doubt by documentary evidence on the threshold of real history. This circumstance likewise renders it clear, why Ophir did not need special explanations in the Old Testament; for the educated world, as it then existed, must have had about the same knowledge of this gigantic enterprise as ours possesses, *e.g.*, of the Congo Free State.

Zimbabye.

Among these ruins, those of Zimbabye or Simbaoe occupy the first place. As already mentioned, they were rediscovered by Mauch in 1871, and he was convinced that he had found historic remnants of the Ophir voyages. He reports on his great discovery as follows: "Zimbabye lies almost due west

of the Portuguese station Sofala or Sofara (Sophara), distant about 160 English miles, and it shows two main groups of ruins. One is on the top of an isolated granite hill, about 400 feet high, the other about half an English mile to the south of the hill, and separated from it by a small silted-up valley. A ringwall, not more than four feet high, partly destroyed and partly buried, runs at some distance from the southern main block round the western foot of the hill; probably it once enclosed it altogether. In the ruin on the hill the outer wall appears to have been erected for defensive purposes, for it is built bodily on the rounded edge of a massive rock, which is 300 feet long by 60 feet high. It runs in a straight line from east to west, is about 120 feet long, 12 feet thick at the bottom, and six feet at the top. Within it an oblong

space is enclosed by somewhat thinner walls. Its western side is rounded off archwise. Here also are visible in the upper, fallen-in parts, balk-like supports of stone, which rise up vertically in the middle of the walls at a distance of eight feet from each other. They seem to have been intended for supports of the wall, which is put together without mortar, and on them leant the individual stones." (Ergaenzungsband zu Petermanns Mitth. VIII, April, 1874, fol. 49.)

Mauch interrogated an old native of priestly family about the object of these buildings; he received the answer, that the ruin in the valley was called the "House of the Great Lady." Also, religious celebrations were still being performed in the ruins on the hill, according to an ancient and curious rite. The sacrifices, in which cattle were killed and their blood sprinkled, appeared,

in Mauch's opinion, to present remarkable resemblances to ancient Jewish rites. "The similarity of those sacrifices to those described by the Jewish religion is obvious. The outlines are there, even if the details leave much to be desired."

"Relying on this, I do not believe I err," says Mauch, "when I assume, that the ruin on the hills represents an imitation of the Solomonic Temple on Mount Moriah, and the ruin on the plain an imitation of the palace in which the Queen of Sheba resided whilst on a visit to Solomon. It may be well supposed, that this Queen who is generally assigned to the realm of myths, became, during her sojourn in Jerusalem, a convert to Judæism, and, conscious of possessing in her kingdom, in the regions of the Sabia river, all the materials and treasures which were used in Solomon's edifices, she resolved to erect similar buildings with the assistance of Phœnician builders. Besides, these ruins correspond well with known Phœnician buildings. Natives and Arabs would have built differently, and the Portuguese already knew of them as ruins" (see fol. 51, A).

When reading these conjectures one is involuntarily reminded of certain rather too positive interpretations by Schliemann of his discoveries in Troy and Pergamon. It is scarcely necessary to tell my experienced readers, that for such positive conclusions and assumptions, as Mauch puts forth, there is no actual basis in the Zimbabye discoveries themselves. No trace of proof has been procured, that the Queen of Sheba had dwelt there, or that an imitation of Solomon's Temple had been intended.

Better founded is the following expo-

sition : " I know, that, through deep studies and energetic diligence, model essays by very learned authorities have been produced, which place Ophir in India, or Arabia, or, who knows where ? Without desiring in any way to disparage these views, I nevertheless believe, that I am bound to give it as my opinion, albeit not first in order, and certainly open to correction, that Ophir is the modern Sofala, or Sofara, as which it is known and so pronounced in the interior ; it is the port, in which the most ancient sailor-nations bartered their own products for the products of the interior. All these products which we know as exported from Ophir are the same, with the exception of peacock feathers, which could still be exported, if better commercial connections were cultivated " (fol. 51, B).

We cannot withhold their scientific justi-

fication from these arguments. Petermann endeavours to support them by producing ancient Portuguese testimony for the same theory. This is very noteworthy, but when we examine Mauch's and Petermann's reasoning without prejudice, we will have again to affirm, that they prove nothing conclusively.

They do contribute towards the probability of the East African theory, but they do no more.

Nor have their conclusions been directly confirmed by subsequent discoveries. The last investigations of the locality have produced no direct evidence, that we have to look in the Zambesi regions for the Ophir of the Solomonic epoch.

**Bent's Explorations.**

It is true, in connection with what we were enabled to establish in the preceding pages, that, which Bent and Swan have

*proved* from the ruins of Mashonaland, becomes at once of really decisive significance in the Ophir question. Let us, therefore, attend, to what they tell us, so that we may be able to form a final judgment on the subject under consideration.

Bent says (The Ruined Cities of Mashonaland, fol. 89) : "I can speak from personal experience of the ruins on the Lundi river ; of those at or near Zimbabwe ; of the chain of fortifications on the Sabi river, including Metemo, Matindela, Chilenga, and Chiburwe, and of the fort on the Mazoe goldfield, which belong all to the same period and race.

"I do not doubt, that, when the country is opened up, many more will be brought to light, proving the large population, which once lived here in occupation of a hostile country for the sake of the gold, which they

got from the quartz-reefs between Zambesi and Limpopo." The descendants of this ancient population we most likely have in the Hottentots of to-day.

"As features of the landscape these ruins are extremely remarkable — old, massive, mysterious, in sharp contrast to the primitive huts of the natives who dwell around them, and to the savage aspect of nature." (fol. 87.)

The general impression produced by the grand ruin of Zimbabye, of which Mauch gave a detailed description, is described by Bent as follows: "This is the grand fortress of Zimbabye," (Bent writes Zimbabwe instead of Zimbabye) "the most mysterious and most complicated erection, which I have ever had the fortune to inspect. One tries in vain to form a clear idea, what it must have looked like before annihilation over-

took it ; with its spiral and well-guarded gates ; its walls bristling with monoliths and round towers ; with its temple adorned with great spectral birds ; its massive cavities, and in the innermost hiding places the business-like, gold-producing smelting furnaces. What was that life like ? Why did the occupiers protect themselves so carefully against attack ? A thousand questions rush upon us, which in vain we endeavour to answer. The only similar sensation, which I have ever experienced, was, when I looked upon the long avenues of Menshirs, near Carnac ; a sensation, which enchants and torments at the same time ; for one experiences the deepest hopelessness to learn all, that one would like to know of the subject. The fortress alone is a sufficient marvel ; but, taken with the vast circular building below, with the numerous ruins, which lie near, and

with the other ruins of the same kind, which lie at a distance, one cannot help to acknowledge the greatness and power of that ancient race ; their great architectural capability and their strategic talent."

What Bent has scientifically verified, is, briefly, the following : These ancient buildings represent throughout a combination of temple and fortress, and were everywhere connected with appliances for the production of gold.

The temples show relations with the worship of the sun. The conical towers or "Phalli," which form the centre, are characteristic ; so are the birds, vultures or eagles of stone, which adorn the outer walls, of which Bent found quite a number in life-size and miniature. These birds resemble the symbol of the Assyrian Astarte, who represents the female element in Creation.

Phalli and these birds together are the personifications of the two divinities, male and female, who represent in primitive Semitic religions the procreative and productive powers of nature, and which were worshipped by the Phœnicians and Sabæans, who were originally settled in Southern Arabia.

Further, these temples are erected in accordance with certain simple arithmetical laws. Also, they have relations with certain astronomical events, in this way, that from their innermost sanctuary the transits of certain stars could be observed. These stars, without exception, belonged to the northern hemisphere. Therefore Bent concludes that the builders were of northern origin. The shape of the great temple of Zimbabye is elliptical and resembles that of a temple of Marib, the ancient Saba, the

capital of the Sabæan Empire ; also that of another temple at Hajar, near the castle of Nakab, both in Southern Arabia. (fol. 93.) The "Phalli," or conical towers, have their analogy in the temple of Byblos, a most ancient sea-town in Phœnicia, between Berytos and Tripolis, which was a seat of the Adonis worship and possessed a temple of Astarte. A representation of that temple, with its typical "Phallus" tower, has been preserved on coins. Similar erections have been discovered in the round temples of Kabira at Hadjar Kem in Malta. The construction of these buildings bears a remarkable resemblance to those of Zimbabye ; the round towers or Nuraghs, in Sardinia, may have had the same significance. It would therefore follow, that the same influence operated in Sardinia as in South Africa. (fol. 100.) "It has been proved that primitive Phœnician influence was at work in Malta and Sardinia."

Bent and Swan found in the neighbourhood of Zimbabye a series of curious stones near an altar, which had evidently formed objects of primitive adoration. Perrot and Chipcez (in their work Geschichte der Kunst in Sardinien, vol. i, 58) write: "We find the worship of the 'bethylae' (holy stones) in every country, to which Phœnician influence has extended." Bent adds (fol. 102): "We will probably be more correct, if we ascribe the worship to a still more ancient Semitic influence, which survived amongst the Phœnicians until more recent times."

Of utensils, the two English scholars found a number of crowbars, shovels, axes, chisels, &c., of very ancient origin; also smelting furnaces of very hard cement with

chimneys; crucibles, polishing sticks and moulds, which exactly resembled those which were found in Falmouth Harbour. The latter were used for tin and are proved to be of Phœnician origin. Traces of gold were still perceptible on the slack of the ingot moulds in South Africa.

Bent discovered only one single trace of an ancient inscription. (fol. 167.) It recalls the pre-Arabian letters of the Sabæans, of which he gives two samples.

I need not follow him in the details of his exposition. Everything he says, confirms, what has been already said. In those ruins beyond Sofala we find buildings of Sabæo-Phœnician origin erected for the production of gold on a vast scale. Their beginnings date from long before the Solomonic epoch in the dimmest primitive phase of human development; probably from a period, when

Sabæans and Phœnicians still formed one nation in South Arabia.

All this renders it clearer, how it could have happened, that Arabia, in itself so poor in gold, could have been proverbial throughout antiquity for its treasures of gold. The Sabæans, who ruled East Africa throughout its entire extent, held the sources of gold south of the Zambesi, and thence they fetched their riches in unmeasurable quantities.

There cannot be really any room left for doubt, that Solomon, who was allied with the Queen of Sheba and with King Hiram of Tyre, sent his vessels on the same errand. Surely he sent them, where his allies had for a long time obtained their gold, where, over large areas, existed mighty establishments of a military and technical nature for the production of gold ; he sent them to Sofala, the ancient country of Ophir. It is not impossible,

that his people advanced as far as Mount Fura to work the goldfields on their own account.

It is my belief, that the documents in the Old Testament, when read without prejudice, refer not to one but to quite a number of Ophir voyages. How could the celebrated passage in 1 *Kings* x, 22, be otherwise interpreted? "For the King had at sea a navy of Tharshish with the navy of Hiram; once in three years came the navy of Tharshish, bringing gold and silver, ivory and apes and peacocks."

If the writer had intended to convey, that only one voyage was made, which it took three years to accomplish, surely he would have written: "After three years the navy returned," &c.

And how do the interpreters explain the passage in 1 *Chronicles* xxix, 4, where David speaks, not Solomon: "I have given three thousand talents (kikkar) of gold, the gold of Ophir, and seven thousand talents (kikkar) of refined silver, to overlay the walls of the houses withal." Such a quantity of gold can certainly not be accounted for by only one Solomonic Ophir voyage. That voyage only brought 420 or 450 kikkar of gold, and David speaks here already of 3,000 kikkar of Ophir gold. It follows with certainty, that David already had relations with Ophir. I refer here once more to a passage in Eupolemos, relating that David had sent miners to Urphe (or Upher), an island rich in gold, who brought gold from there to Judæa. The term "island" need not trouble us, for a continent that had to be reached by sea, was an island to the ancients. Does not this passage complete the other in *Chronicles*, and are we not right to interpret Urphe as Ophir? It is strange, that this point has

been missed by all the expounders of the Ophir question.

As far as I am concerned, I hold, that an unprejudiced consideration of all the different accounts must bring us to the following result: the Ophir enterprise did not mean a single voyage, but must be understood as an enduring activity, directed by Jewish policy in Africa towards a steady production of gold. It has been shewn that this policy was followed by David and Solomon, and that a hundred years after the latter's death, Jehoshaphat vainly endeavoured to take it up again. During Solomon's reign the results of the gold production were once in three years brought to Jerusalem.

* * *

**Final result.** I have finished my investigation. It has a twofold result. Firstly: the ancient name Ophir has survived up to our own times in its Latin form, Africa. Secondly: it is conclusively established, that the part of Africa, to which Solomon's gold fleet sailed, must be looked for in the primitive Sabæo-Judæan gold country beyond Sofala. That it should have been specifically known by the name of the entire continent is a matter, for which analogies can be found even in our own times. The Romans called their province in Asia by the name borne by that entire part of the world.

The United States of North America are, in common parlance, called America by their citizens, and in the eighties the popular idea of Africa was expressed by the Germans as Cameroon. The ancient Jew thought of Ophir, in the first place, as of the country, with which he stood in immediate practical relations; that is to say, he meant by Ophir the regions of the Zambesi beyond Sofala.

The chain of events, which reaches so very far back, is now completed without a break up to our present times. For we know that there, where Sabæans, Phœnicians and Jews unearthed their vast stores of gold, British Colonial enterprise has proclaimed the modern gold-fields of Mashona and Manica-land. One thing remains yet to be done: the thorough exploration of ancient Ophir. There is no doubt whatever, that very many and very important things still remain hidden under the debris of that gigantic enterprise in South Africa.

Whoever leads the exploration, must command that wide range of knowledge, which alone can decipher such phases in the history of the human race. And above all must he possess that rare power of perception, which enabled a Cuvier to reconstruct the entire skeleton of an antediluvian animal

organism from the single specimen of its bones, which chance preserved through countless ages for the enrichment of our knowledge of the times, when man was not yet created to chronicle their progress.

As I am on the point of leading myself an expedition for mainly practical purposes into these very districts, I hope I shall be able to contribute somewhat to the further solution of the problems, which have been set forth in this research. I shall take all the necessary instruments with me for carrying out a careful exploration of the ruins, which Karl Mauch was the first to discover and Bent has described in his book, and into the secrets of which we have to penetrate in order to finally clear up the mysteries of that most ancient epoch of human striving after the precious metal.